BRITAIN
TODAY

FARMING

ROY WOODCOCK

WAYLAND

TITLES IN THE SERIES
CITIES
THE COUNTRYSIDE
THE ENVIRONMENT
FARMING
INDUSTRY
TOURISM

Series editor: Sarah Doughty
Book editor: Ruth James
Series designer: John Christopher
Production controller: Carol Titchener
Consultant: Gordon Hickman, arable and livestock specialist
at ADAS

First published in 1995 by Wayland (Publishers) Limited
61 Western Road, Hove, East Sussex BN3 1JD, England.

British Library Cataloguing in Publication Data
Woodcock, Roy
Farming. – (Britain Today Series)
I. Title II. Series
630.941

ISBN 0-7502-1534-8

Picture Acknowledgements
Bruce Coleman Ltd 37; Ecoscene 8, 27 ,42 (bottom), 42–3 (top), 44–5;
Eye Ubiquitous 9, 17, 18–19, 25, 36, 38; Holt Studios International 6; Impact 29
© Christopher Cormack; Milk Marque 20; F Motisi 14–15; National Dairy
Council 21; British Sugar Bureau 16; Environmental Picture Library 22, 32–3,
34 (top), 35; Simon Warner 31; The Rural History Centre 12;
Tony Stone *cover*, 26, 38–9; Zefa Picture Library 4, 11.
The remaining pictures are from the Wayland Picture Library.

Typeset by White Design
Printed and bound in England
by B.P.C. Paulton Books

CONTENTS

INTRODUCTION

Farming has been practised for thousands of years in Britain. Generations of farmers have made the best use of the conditions of a particular area and adapted the land for cultivation or livestock. Traditionally, different types of farming have taken place in different areas of Britain, and the way farming has developed has depended on the soil, climate and **altitude** of the particular region.

▼ Heavy rainfall keeps the countryside green and maintains rich grassland in western Britain.

Some areas of Britain, such as the Highlands of Scotland, have poor soils and harsh weather conditions. Vast areas of the Highlands, particularly at the higher altitudes, are largely **uncultivable** but produce some rough grazing for sheep. On slightly lower ground, crops can be grown. However, the cool, moist climate of the north of Britain allows only barley and oats to grow well in the short five-month growing season. Cattle are kept in fields on the lower ground in sheltered areas.

The weather systems coming in from the Atlantic bring rain to the lowland areas, particularly on the western side of Britain. These areas are important for **pastoral** farming which can be based on dairying, beef or sheep. In the south of Britain, warm temperatures mean a long growing season for crops – eight months or more. Cereals ripen more easily in southern Britain because of the warm conditions. Many mixed farms are found in the south, with **arable land** for growing crops and grassland for cattle. A mixture of arable farming and livestock is also found in many other parts of Britain, including the Midlands, the North-East, the North-West and parts of Scotland.

Types of farming in Britain. ▶

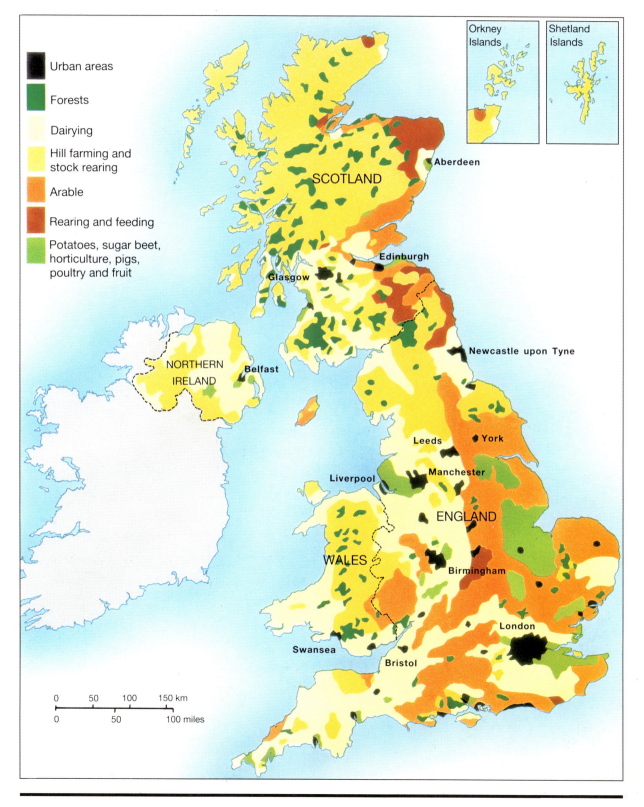

Urban areas

Forests

Dairying

Hill farming and
stock rearing

Arable

Rearing and feeding

Potatoes, sugar beet,
horticulture, pigs,
poultry and fruit

Orkney
Islands

Shetland
Islands

SCOTLAND

Aberdeen

Edinburgh

Glasgow

NORTHERN
IRELAND

Belfast

Newcastle upon Tyne

Leeds

York

Liverpool

Manchester

ENGLAND

WALES

Birmingham

London

Swansea

Bristol

| 0 | 50 | 100 | 150 km |

| 0 | 50 | 100 miles |

Mixed farming has given way to more specialized farming in some parts of the country. In East Anglia, parts of Yorkshire, Lincolnshire and eastern Scotland, farmers have decided to concentrate mainly on crops. East Anglia is a lowland area and a few places, such as the Fens, have rich **peat** soils. The dry, sunny summers help cereals to ripen and the removal of hedges has made the area ideal for cultivating crops which need large, flat fields in order to use machinery such as combine harvesters.

Market gardening is usually carried out in large specialist units. Market gardens are common in the central counties of England, in areas of flat land and **fertile** soil. Market gardens are usually located near a market centre where produce such as potatoes, tomatoes and cabbages can be quickly transported and sold.

Smallholdings still exist, but the majority of vegetables are grown on a large scale. In the south-east of England, the mild climate has allowed orchards of apples, pears and cherries to grow successfully.

Although the relationship between soils, climate and landscape is important, all farms are different since farmers have their own preferences about the types of crops and animals that are kept. Sometimes this depends on personal choice and the amount of money a farmer wants to put into the farm.

Farms are businesses so the farmer has to make decisions about how **intensive** a farm needs to be in order to make a profit. Farmers also have to abide by the rules and regulations laid down by the European Union (EU), which provides **subsidies** and grants to farmers.

◀ Vegetables are grown intensively and generally close to main towns so that they can be sold while still fresh.

Farming is a business ▶ and must be efficient and highly organized. Machines have replaced farm labourers for much of the work.

THE COMMON AGRICULTURAL POLICY

The European Economic Community (EEC) was founded in 1957, with six countries as members. In 1994, the EEC became the EU. Following the joining of three further countries in 1995 the EU is now made up of fifteen member states. The EU is responsible for the Common Agricultural Policy (CAP), which is a set of rules that all members have to abide by. The aims of the CAP were laid down in 1957 and were as follows:

• to produce more food;
• to make sure farmers have a fair standard of living;
• to stabilize markets to provide a steady supply of food;
• to make sure food is reasonably priced.

In order to achieve these aims, the CAP has a set of guide prices throughout the EU. Farmers should receive the same price for the sale of their crops and animals. To encourage people to buy British food in the UK, taxes are placed on goods from non-EU countries. This helps British farmers in the agricultural market. They are also protected from prices going too low, which would mean that they would receive little income from their produce.

The fixing of prices affects what farmers grow. In the 1970s for example, when farmers received a good price for oilseed rape many farmers began to grow this crop for its oil. Because of the policy of fixing prices, farmers in the past were encouraged to grow too much, causing **surpluses**. These surpluses could not be sold and were stored in Britain and other European countries. They were called food mountains – for example, there are beef and grain mountains.

▲ Fruit and vegetables are grown in many parts of Britain. British farmers have to compete with producers in warmer countries of the EU.

▲ Large areas of southern Britain turn yellow in early summer as fields of oilseed rape begin to flower.

In 1993 the General Agreement on Tariffs and Trade (GATT) – a worldwide agreement intended to make international trade easier and fairer – meant that the CAP had to be modified. In particular, GATT requires that surpluses be reduced and that financial aid to farmers does not distort international markets.

In order to comply with the needs of GATT the EU introduced a series of measures, one of which was the policy of set-aside. Set-aside is where farmers are encouraged by subsidies or financial aid to put aside land that would usually be used for growing arable crops. They are encouraged to grow an alternative crop, or simply maintain their set-aside land in good condition. In order to qualify for Arable Area Payments farmers must set aside a minimum of 12 per cent of their arable land.

Other schemes that have been introduced, such as Countryside Stewardship and Environmentally Sensitive Areas schemes, allow farmers to receive money for managing special landscape areas such as chalklands, heathland and woodland.

MANAGEMENT

A farm is a business, and most of Britain's farmers are **commercial** farmers who produce food or goods for sale. To produce goods effectively, a farmer has to manage his or her business in a particular way. Each farmer relies on a system, made up of inputs and outputs. To get something out of the farm, the farmer has to put certain things in. Some inputs are natural, such as soil, rain and sunshine. Others can be seen as investments – the farmer puts in money, machines, labour, fertilizer and seeds. By doing this, the farmer hopes to achieve results or outputs – the growth of cereals and root crops or animals that can be sold.

Inputs and outputs are different on farms of different types and sizes. For a cereal farmer in Norfolk, for example, the major inputs are sunshine, rain, soil, fertilizer and several machines including combine harvesters to harvest enormous fields of crops. Compare this to a small dairy farmer in Dorset or Devon where the major inputs are sunshine, rain and nitrogen to improve pastures. The dairy farmer may also have to maintain hedges which form boundaries to the fields, and make hay and **silage** for feeding the cattle in winter.

▼ The farmer's input.

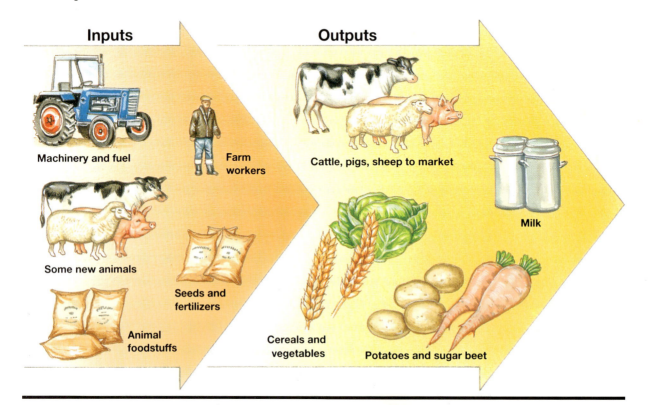

Inputs

Machinery and fuel

Farm workers

Some new animals

Seeds and fertilizers

Animal foodstuffs

Outputs

Cattle, pigs, sheep to market

Milk

Cereals and vegetables

Potatoes and sugar beet

▲ Manure is spread over fields of pasture to improve grass growth and increase productivity, making farming more intensive.

Farms vary as to how much produce and income they generate. On some farms, there is only a small amount of produce and income from each **hectare** of land. This is called an **extensive farm**. On intensive farms each hectare of land is made to produce a high income. Intensive farms usually require high inputs.

PROGRESS AND CHANGE

Farming is constantly changing, as new machines are developed by engineers and new seeds are developed by plant breeders. In the early eighteenth century, many important developments took place that helped create the farming we are familiar with today.

One of these developments was the horse-drawn seed drill invented by Jethro Tull. His seed drill sowed seeds in neat rows which was better than scattering the seeds by hand. He also used a horse-drawn **hoe** to loosen the soil around his crops, allowing water to reach the plant roots more easily.

Early machines revolutionized ▶ farming methods. This is Jethro Tull's horse-drawn seed drill which was invented in about 1701.

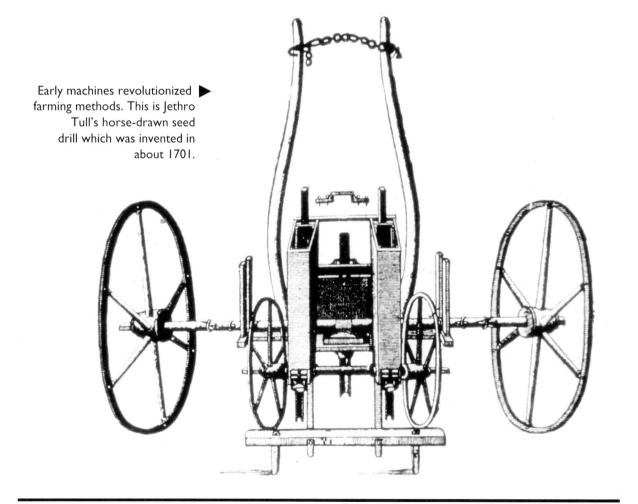

A FOUR- FIELD CROP ROTATION

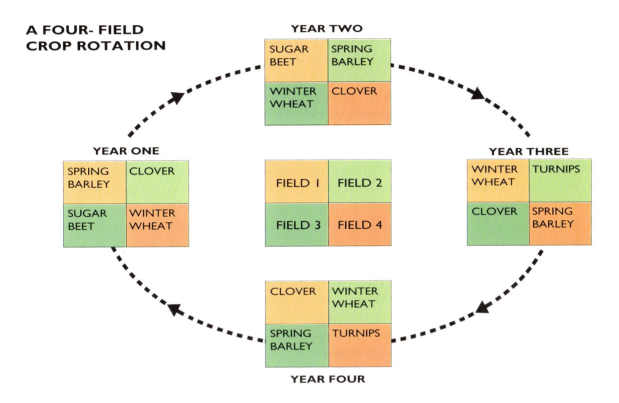

YEAR TWO

SUGAR BEET	SPRING BARLEY
WINTER WHEAT	CLOVER

YEAR ONE

SPRING BARLEY	CLOVER
SUGAR BEET	WINTER WHEAT

FIELD 1	FIELD 2
FIELD 3	FIELD 4

YEAR THREE

WINTER WHEAT	TURNIPS
CLOVER	SPRING BARLEY

CLOVER	WINTER WHEAT
SPRING BARLEY	TURNIPS

YEAR FOUR

Another important development was made by 'Turnip' Townshend, who started the Norfolk system of rotation. Instead of having a **fallow** year, **fodder** crops were grown in rotation with other crops, often cereals. Rotating crops kept the soil healthy and helped produce more crops. Rotation is still used today and works best when one crop provides nutrients or conditions required by another crop. For example, grasses and beans may help cereals to grow by providing nitrogen to improve the soil.

Farm machinery, especially tractors and combine harvesters, have improved the efficiency of work done on farms. Machines have also allowed the farmer to manage with fewer workers. The tractor is the most important piece of equipment because many specialized machines can be attached to it. There is machinery for preparing the soil, sowing seeds, spreading fertilizer and harvesting crops.

Preparing the soil is done using tillage machinery. A plough turns the soil over and starts the process of breaking it up. Harrows and cultivators are then used to break it down further. Mechanical seed drills put seeds into the ground, spreaders are used to add fertilizer, and sprayers add chemicals to control pests and diseases. Finally, a variety of harvesting machines are used to gather crops when they are ready. Forage harvesters gather grass and maize for silage, and combine harvesters gather grain crops.

ARABLE FARMING

Arable land is used for growing crops. These crops may be cereal crops such as wheat, barley and oats, or root crops such as potatoes and sugar beet. The two most important grains are wheat and barley, but in wetter areas oats are grown. Cereals need summer temperatures of 17 °C in July and August for them to ripen.

FACT BOX

CEREAL CROPS IN BRITAIN 1994
(area per thousand hectares)

Total cereal area	3,000+
Wheat	1,800
Barley	1,100
Oats	109
Others	16

FACT BOX

AREA DEVOTED TO CROPS 1994
(area per thousand hectares)

◻	Wheat	40.4%
◻	Barley	24.7%
◻	Oil seed	9%
◻	Sugar beet	4.4%
◻	Peas/beans	5.1%
◻	Horticultural	4.2%
◻	Other	12.2%

In East Anglia, 50 per cent of the land is used to produce cereals. The land is flat, which makes it easier to use farm machinery. To create large fields and make tilling and harvesting easier still, farmers have removed many of the hedges that existed before. However, hedges form natural barriers that help prevent wind from blowing the soil away. Removal of the hedges has allowed **soil erosion** to take place.

Many farmers grow large areas of one crop so that they get full use out of their machinery. Sometimes the same crop, such as wheat or barley, can be grown for several years. This practice is called monoculture.

A combine harvester cutting the ripe wheat ▶
while the weather is dry.

Unfortunately, the practice of monoculture means that the same nutrients are taken from the soil year after year, and the soil becomes exhausted unless large quantities of fertilizer are put back into the soil to keep it fertile.

The use of large amounts of fertilizers has become a sensitive issue. This is because they may be washed through the soil and into the rivers which then become polluted. One solution to this problem is the use of crop rotation. The usual rotation contains cereals for two or three years, then a root crop or **leguminous** crop such as beans, peas, clover or alfalfa, to help restore nutrients to the soil.

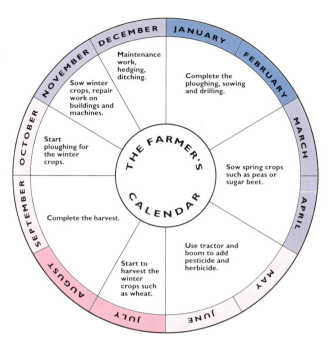

THE FARMER'S CALENDAR

JANUARY — Maintenance work, hedging, ditching.

FEBRUARY — Complete the ploughing, sowing and drilling.

MARCH — Sow spring crops such as peas or sugar beet.

APRIL / MAY — Use tractor and boom to add pesticide and herbicide.

JUNE / JULY — Start to harvest the winter crops such as wheat.

AUGUST — Complete the harvest.

SEPTEMBER — Start ploughing for the winter crops.

OCTOBER / NOVEMBER — Sow winter crops, repair work on buildings and machines.

▲ Once harvested, the sugar beet has to be rushed to the refinery before any of the sugar content is lost from the crop.

JERSEY

Jersey is one of the Channel Islands, which lie off the coast of Brittany. It is well known for its early crop of potatoes. Jersey Royals were developed as a distinctive variety of potatoes in the 1880s and have been grown ever since.

Jersey Royals grow on land enriched with seaweed. About 200 growers plant over 3,000 hectares, producing over 40,000 tonnes per year. The potatoes are grown in small fields near the sea, protected by polythene sheets in February and March. Some Jersey Royals are grown indoors and they are harvested from March to April, by which time the outdoor crop is about ready.

◄ Small-scale intensive farming of potatoes requires human labour, but earns a high return per hectare.

The main root crops grown on a farm are sugar beet and potatoes. Sugar beet is a fairly recent crop to be grown on British farms. It grows well on most soils and is usually rotated with cereals. It is usually sown in March, and harvested in late autumn by a machine that lifts the beet from the ground, tops it and feeds it into a trailer. Sugar beet is always grown near a refinery. When the sugar has been extracted from the root at the factory, the waste is used as animal feed.

Potatoes are usually sown as small **tubers** in spring, often in shallow ditches with the soil pushed in on top, partly to protect them from frost. Because potatoes are prone to infection, as they grow they need to be sprayed with **fungicide**. Mechanical harvesters are usually used for digging the soil, but care is needed not to damage the potatoes as they often need to be stored. Early potatoes are collected as soon as the soil is dry enough, usually by handpickers.

DAIRY FARMING

On a dairy farm a farmer breeds and raises cows for milk production. The products that can be made from milk include butter, cream, cheese, ice-cream, dried milk, skimmed milk and yoghurt.

Dairy cows are farmed in areas where there is plenty of pasture and where the grass grows quickly. Some of the main dairy farming areas are in Cornwall, Devon, Somerset, Cheshire and south-west Scotland. Even though it is more expensive to keep cattle in the drier, cooler climate of eastern England, dairy farming takes place near many large towns and cities because it is easy for the farmers to sell their milk at the many local markets. Milk needs to be moved quickly so that it does not go sour.

The dairy farmer has to work every day of the year. The cows are milked regularly. This keeps their milk flowing for up to about 300 days after calving. Milking usually takes place early in the morning and again in the evening. Milking on many dairy farms today is mechanized and carried out under very

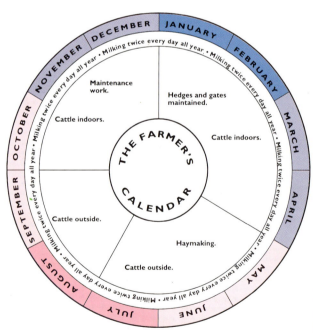

▲ Although milking is mechanized and computerized, farmers still have to get cows into the shed twice a day, every day of the year.

◄ Cattle spend much of each day lying down and chewing to break down the cellulose in grass and convert it into milk.

clean conditions. Each cow is on a computer file and the amount of milk produced is recorded. While being milked, they are given feed high in nutrients to help maintain a high milk output.

Dairy cattle today have been bred to produce large quantities of milk. Black and white Friesians and Holstein, and the brown and white Ayrshire cattle are good milk producers. The Jersey or Guernsey herds do not produce such large quantities of milk, but their milk is richer.

COLLECTION AND DISTRIBUTION OF MILK

Milk is stored in storage tanks on the farm until it is transported via tanker lorry to the nearest factory, where bottling, cheese- or butter-making takes place.

Up until the end of 1994, collection and delivery of milk was carried out by a national organization called the Milk Marketing Board. Today, Milk Marque, a co-operative owned by its farmer members, and other smaller independent companies have taken over the responsibilities of marketing, selling and distribution of milk. Milk has entered a competitive or 'free' market, with many companies competing to sell and distribute the farmers' milk.

The price that farmers pay for collection, marketing and distribution of their milk depends on how many companies enter the market and how many farmers need this service. These companies say that farmers will be able to pick and choose whom they wish to transport their milk to the dairy and that they will receive a fair price from the dairy company for their milk.

Once the milk has been transported to the dairy, it has to be pasteurized. Pasteurization of milk is a heat-treatment process that destroys micro-organisms found in milk. Pasteurization also prolongs the amount of time milk can be stored.

Modern pasteurizing equipment holds the milk at high temperatures, from 72 °C for sixteen seconds and up to 85 °C for a brief moment. Once the milk has been pasteurized, the bottling or packaging machine fills and seals the milk in retail containers.

◄ Bottling of milk is done under sterile conditions.

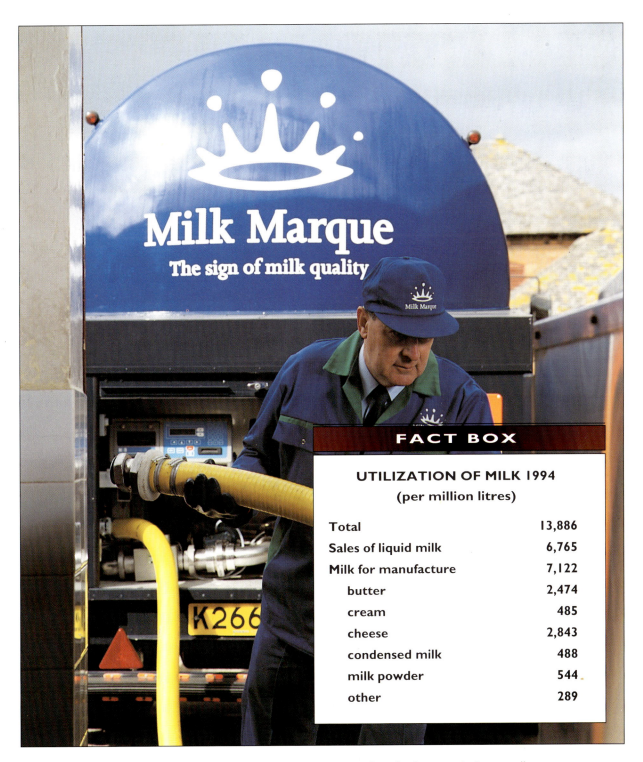

FACT BOX

UTILIZATION OF MILK 1994
(per million litres)

Total	13,886
Sales of liquid milk	6,765
Milk for manufacture	7,122
butter	2,474
cream	485
cheese	2,843
condensed milk	488
milk powder	544
other	289

▲ Milk Marque and several smaller companies send tankers round to the farms each day to collect the milk. Much milk is bought by the large supermarkets.

BEEF FARMING

On a beef farm, cattle are raised and bred for meat. In Britain, a beef farm is the most common type of farm. There are over 11 million farms for cattle, many of which also have sheep. Some of the farms are on the lower slopes of hills and mountains in Scotland, Wales, Northern Ireland, parts of the Midlands and Devon. Other beef cattle are reared on lowland farms in Leicestershire, Norfolk and as far north as Aberdeen. These include the Aberdeen Angus bred in the north-east of Scotland, and the Hereford, bred on farms along the Welsh border.

▼ Breeders produce top-quality bulls like this Aberdeen Angus to maintain or improve herds of beef cattle.

▲ Cattle need additional feed in winter when there is little grass for them to feed on.

Much of the beef produced in Britain actually comes from the dairy herd. Calves which are a mixture or cross of both dairy and beef cattle are bred on the dairy farm and sold to beef producers who then raise them for their beef. Once they reach one year of age many cattle are sold as store cattle. Store cattle fend for themselves by finding food from grazing on the poorer pastures of the hills. After about a year they are moved to richer lowland pastures or to different lowland farms. Here they are given feed high in nutrients, such as sugar beet, turnips and cattle cake to improve the quality and quantity of their meat before being sold for slaughter. Once they have been fattened up, many cattle are taken to market to be sold.

Markets take place every week for the sale of cattle and sheep and are very busy at the end of the summer when lambs and calves are sold. One of the oldest cattle markets is Welshpool in mid-Wales, which has been taking place for over 100 years. During each year 40,000 cattle and 600,000 sheep are sold there.

SHEEP FARMING

▲ Sheep dogs are important for rounding up the sheep to move them to richer pasture, or for gathering them for dipping and shearing.

Sheep are kept for their meat, wool and in some cases their milk. There are many kinds of sheep – over fifty different breeds in Britain alone. A number have been bred for the quality of their meat. Such breeds are generally raised on lowland farms, where it is possible to keep up to twenty sheep to a hectare. Hardier breeds are kept on hill farms. Here the grazing is poor and as few as two sheep per hectare may be kept.

Consequently, hill farms tend to be very large. Some of these farms in the mountainous areas of Wales, Scotland and the Pennines have 6,000 or more hectares of land.

About 500 years ago, wool was the most important sheep product. Today, many sheep farmers get most of their income from selling meat, usually in the form of lamb.

However, wool remains an important product. After a sheep is shorn, the fleece is graded, washed and spun before being used to manufacture clothing and carpets.

On hill farms, lambing usually takes place in April and May. During the winter, the shepherd or shepherdess may have tended the flock on the hillsides, or he or she may have brought them down to spend the winter in the valleys. On lowland farms, sheep may lamb at almost any time of year. Usually, lambs start to appear in November, with a peak period in February, March and April. Lambing is the busiest time of year and on many farms it takes place in covered sheds.

CASE STUDY

LITTLE LANGDALE

Much of the Lake District is used for sheep farming. The farms are situated in the valleys, but the sheep – mostly Herdwicks – live on the hillsides for several months of the year. The fields nearest to the farm, called inbyes, provide rich pasture and areas in which to grow crops of fodder and vegetables. The soils are deeper and more fertile than on the hillsides and the meadows nearest the farm are used for lambing. The fields on the lower slopes, called intakes, are also used for grazing and for hay. During the summer, the higher fell land is used for grazing the sheep.

Heavy rainfall in the Lake District helps ▶ to ensure rich green pastures for sheep throughout the summer months.

MIXED FARMING

Mixed farming means growing crops and rearing animals on the same farm. In the past, most farmers were mixed farmers, but commercial farming has become more specialized. It is easier for the farmer to buy the most suitable machines and grow just one or two crops. When mixed farming stopped, crop rotations were not used, so farmers began to put more fertilizers on the soil to keep it fertile. Some of the largest farms practised monoculture, and wheat and barley were the most common crops.

In recent years there has been a move back to more mixed farming. It is less risky than growing only one or two crops. Some of today's farmers who manage mixed farms believe in **organic** farming and so do not use any chemicals. The reasons for this change to mixed farming is because of over-production in Europe and because the soil has been getting less fertile despite the addition of large quantities of fertilizer.

Since the Second World War, large areas of Salisbury Plain and the chalkland areas of Wiltshire, Hampshire and Berkshire have been used mainly for growing cereal crops. In the last few years there have been many types of crops planted, with more fields of

oilseed rape, vegetables and grassland. In the grassland areas flocks of sheep and herds of beef cattle are reared, and they help to make the soil fertile because their dung improves the quality of the soil.

In the wetter parts of western Britain there have always been more mixed farms than the drier eastern parts of Britain. In Northern Ireland, weather and soil conditions mean crops can be grown and animals reared too. Near to Lough Neagh crops include fruit, vegetables and some cereals, and there are also dairy cattle. Food is produced for local markets and for sending to Belfast.

Fresh fruit and vegetables can be bought at ▶ street markets, roadside stalls and farm shops.

▲ A combination of crops and animals can help maintain soil fertility as well as give a variety of products for the farmer to sell.

KENT

Kent has long been famous for its hop fields and fruit orchards. As cereal growing has become more profitable, wheat and barley fields have increased during the last twenty or thirty years. Hop fields have disappeared, and oast houses that were used for drying the hops have been converted into houses.

There are also fewer fruit orchards because cereals have become more profitable and because fruit grown in sunnier places further south in Europe can be produced more cheaply than in Britain. Recently there has been a move back to mixed farming throughout Kent.

Fruit picking can be mechanized, but ▶ some is still done by hand.

Many small farms have been set up in derelict and neglected parts of cities. These tiny mixed farms provide a break from city life, as well as providing somewhere to learn about plants and animals. A city farm may contain small numbers of goats, cows, sheep, pigs, rabbits, ducks, geese and hens. There is often a grazing area and vegetables and flowers are grown.

To set up a city farm takes time, effort and money. The first requirement is housing or shelter for the animals and, if necessary, greenhouses for crops. Costs can be kept down by using animal dung as manure. Before planting can take place the soil will probably need to be enriched with manure and fertilizer. There must also be storage space for equipment and feedstuffs.

Maintaining a city farm also requires a great deal of effort. Animals need to be looked after, and in the confines of a small city farm will probably need feeding regularly. The health of the animals must be maintained and, as on all farms, this involves such things as routine checks, vaccination and foot trimming. Tasks such as cleaning out animal pens and watering and weeding crops have to be done regularly.

▼ City farms provide children who live in cities the chance to help feed and look after animals.

Costs can be kept down by ensuring that all the animal waste is used as manure, and by supplementing the food the animals get with human food scraps. However, city farms are, by their nature, not easy to keep going – true farming needs the inputs that only the countryside can provide. However, the effort involved in a well-maintained city farm is worthwhile. It provides many people who would otherwise never come into contact with animals the chance to see and touch them, and they can be a useful source of fresh food for the local community.

HILL FARMING

Hill farming takes place in all the main upland areas of Britain, especially in Scotland, Wales and parts of western Ireland. Smaller areas are found in the Pennines and Lake District, as well as Dartmoor and Exmoor in the south-west of Britain.

Conditions are often harsh in these areas and this makes farming difficult. Winters are cold and long, with snow covering the ground for several weeks. Rainfall is heavy, with up to 2,000 mm or more in a year, and the soils are generally shallow and infertile. Hill farmers do not grow many crops because conditions are not suitable, but they

do rear animals, especially sheep. Dairy cattle are reared in the valleys near farm buildings, so that they can easily be brought in for milking. Beef cattle graze on the hillsides during the summer, but are brought nearer the farms in winter, when they are given other feed as well as grass.

Crofting is found mainly in the Hebrides, western Scotland and the Shetland Islands. In the past, crofts were made up of strips of common land which were divided amongst crofters every year. Now crofters have their own arable land fenced in, although they share common grazing land and farm

◄ Cattle graze in the valleys for part of the year, but move up on to the hillside for the summer months.

The wet climate of ► the Outer Hebrides is good for growing grass, and several crops of hay for winter feed can be cut each year.

machinery. A typical croft consists of just a few hectares of land, a handful of sheep, a cow and enough crops (mainly potatoes and oats) to add to the crofter's diet and income.

Many of the crofters have a simple lifestyle – in parts of the Shetlands and Northern Scotland their cottages are heated by burning peat. Cloth, known as Harris Tweed, is woven in a traditional way on handlooms at home. Many of the crofts are too small to provide a living for a whole family so many crofters seek work in the North Sea oil industry, the Forestry Commission or abroad. Crofting as a way of life is becoming rare as young people move away to find jobs. Those who are left behind are anxious to maintain a way of life that is not found elsewhere.

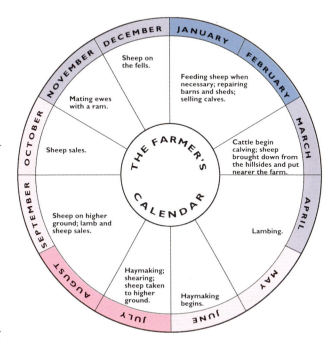

THE FARMER'S CALENDAR

JANUARY
FEBRUARY — Feeding sheep when necessary; repairing barns and sheds; selling calves.
MARCH — Cattle begin calving; sheep brought down from the hillsides and put nearer the farm.
APRIL — Lambing.
MAY
JUNE — Haymaking begins.
JULY — Haymaking; shearing; sheep taken to higher ground.
AUGUST
SEPTEMBER — Sheep on higher ground; lamb and sheep sales.
OCTOBER — Sheep sales.
NOVEMBER — Mating ewes with a ram.
DECEMBER — Sheep on the fells.

MARKET GARDENING

Many vegetables come from mixed farms, but they also come from market gardens, which also sometimes produce fruit and flowers. Some vegetables are grown outside, such as cauliflowers, cabbages, leeks, beans and courgettes. Vegetables that require a warmer temperature, such as peppers, tomatoes and cucumbers are grown in greenhouses in artificial climates. Greenhouses can also be used to grow flowers.

Greenhouses that are used for growing lettuce, tomatoes and cucumbers are heated by electricity or hot water pipes. Seedlings are often raised in cold frames built of metal or brick and covered by a glass frame called a light. The glass allows just enough heat and light for the seeds to **germinate** early. Some seeds and plants that are grown indoors are later planted outside.

Many of these crops grown in greenhouses or outside in small plots are grown intensively a few hectares of land are used to produce a large quantity of crops.

Greenhouses allow crops to ripen much ▶ earlier than is possible outside.

FACT BOX

MARKET GARDENING IN BRITAIN 1994
(area per thousand hectares)

Vegetables in the open	127
Orchard fruit	32
Flowers, bulbs	13
Soft fruit	32
Glasshouse	2

The farms are small – 34 per cent are less than 1 hectare in size and 67 per cent less than 5 hectares. Fields used for strawberries will produce 10-12 tonnes of fruit per hectare, depending on the type of plant and the weather conditions. Some farmers grow crops to be picked by the customers themselves. These 'pick your own' fruit farms are busy during the summer months.

▲ Early outdoor tomatoes are grown in parts of Britain with mild weather conditions such as the south-west.

▲ Many farmers reduce their labour costs by growing fields of 'pick your own' crops, which are popular with families at weekends in the summer.

Early crops are very popular and people are willing to pay higher prices for produce during the winter. British farmers have to compete with crops imported from the Canary Islands or southern Spain. The crops grow best in parts of the country where the soil and climate are favourable. Early potatoes and carrots come from the south-west of England, celery and onions from the Fens, and Brussels sprouts from Bedfordshire. Western Cornwall, the Isles of Scilly and the Channel Islands are all important areas for fruit and vegetables because they have mild winters. Early crops of flowers and vegetables are grown in the Channel Islands, south-west England and the Pembroke area of south-west Wales.

THE ISLES OF SCILLY

The Isles of Scilly are situated off the south-west coast of Britain. They are surrounded by the warm waters of the North Atlantic Drift, so the air temperatures are warm but it is often windy. Farms are often surrounded by dry stone walls which provide shelter from the wind.

The major crop grown is flowers, but potatoes are also grown and harvested from April to June. The Isles of Scilly produce about 50,000 boxes of flowers, which are mainly narcissi. The mild winters mean that harvesting can take place from October until the end of March. Pinks as well as narcissi are grown and are harvested all year. The flowers are transported by boat, helicopter and aeroplane to other European countries.

▼ Mild winters allow early daffodils to be grown in small fields on the island of Tresco.

INTENSIVE FARMING

Intensive farming involves growing crops or keeping animals in a relatively small area. This is to reduce costs and maximize the output of the farm. Intensively grown crops such as sugar beet are generally maintained by adding large amounts of fertilizer to the soil. Animals farmed intensively are generally kept indoors and fed controlled amounts of food.

The main advantage of intensive systems is that animals which are kept warm and well-fed indoors usually produce more or fatten more quickly than they would outside. Intensively farmed animals include chickens (both for eggs and meat), fish, pigs, veal calves and beef cattle.

▼ Outdoor farming can be intensive, as on sugar beet farms in East Anglia.

▲ Pig farming is intensive and can give a large return per hectare.

It is widely believed that all 'factory-farmed' animals suffer because they are deprived of their natural surroundings. However, we cannot be certain of this. Animals that are suffering do not thrive, whereas most intensively farmed animals do. In fact, there is little doubt that, given the choice, animals often prefer to be where food and warmth are provided. For example, free-range chickens are in most cases kept just as intensively as **perchery** birds. Despite the fact that they can go outside, many of them never leave the the shed. The hens seem to prefer the dark conditions of the shed.

Intensively farmed food is often thought to be less nutritious than extensively farmed produce. In some cases this may be so, as there is evidence that meat quality is affected by intensive farming. Keeping animals in crowded conditions does result in some stress. Meat produced intensively often appears to have less taste and may have a higher proportion of fat than meat reared less intensively.

Apart from these considerations, intensive farming has become controversial because it is seen as cruel. Practices such as keeping chickens in battery cages, veal calves in crates and pregnant sows in **farrowing pens** have all been widely criticized.

▲ Calves are intensively farmed, but in Britain they are not allowed to be reared in crates.

Because of the stress caused to the animals many farmers have moved away from intensive schemes of this type. At the same time, it is unlikely that animals have the same feelings as humans. Some would argue that they appear content to live in conditions that we would find intolerable. The important thing is to ensure that animals

▲ Battery hens produce a large number of eggs in a limited space. Some people consider this to be very cruel.

suffer as little as possible. Some practices may still need to be changed. Rearing veal calves in crates has been outlawed in Britain, but it still happens elsewhere in Europe. Most of the eggs eaten come from battery-farmed chickens and nearly all chicken meat is raised in **broiler houses** because intensively farmed food is cheaper to produce than extensively farmed food. In the end, some people say it is up to us, the consumers, to decide what happens. If we decide to pay more for food, then more food could be produced less intensively.

FISH FARMING

▲ As the fish grow they are transported to larger tanks.

Fish farming is an intensive type of farming. It is about rearing fish under controlled conditions. Trout and salmon are commonly farmed, particularly rainbow trout. Some shellfish, such as mussels, shrimps and oysters are also kept on farms.

Most fish are reared in ponds or tanks near a stream or river which provides a fresh supply of water. After the eggs have been **incubated** they are moved to a water tank. As they grow, they are transported from one tank to another. The farmer's main expenses are for the tanks and the pumping equipment used to move water from one tank to another. The fish require a fish feed (usually pellets of protein) little and often and a flow of water to obtain oxygen which the fish need to breathe. They take the oxygen from the water as it passes through their gills.

▼ Some fish farms are located in inlets of the sea where sea water brings in fresh supplies of food every high tide.

In recent years, fish farms have been set up in Scotland – particularly in **lochs** or inlets of the west coast and the Hebrides. Since salmon normally spend part of their life cycle in the sea and part in fresh water, some Scottish farmers have both fresh and salt water sites to copy natural conditions more closely. Because these types of farms can take advantage of sea water they are less expensive to set up than fresh water farms.

ORGANIC FARMING

Modern farming methods are causing people concern for various reasons. The use of fertilizers and pesticides, and the production of large amounts of animal waste, can all result in pollution. Many pesticides are known to leave traces of chemicals in food products. Food surpluses are seen as being unnecessarily wasteful.

To many people, the best solution to all these problems is the system of agriculture known as organic farming. Organic farmers do not use any artificial fertilizers or chemical pesticides on their crops. Animals are raised on organically grown fodder, and only a small quantity of modern medicines are used to treat and prevent disease. Some organic farmers use **homoeopathic** remedies only in treating disease.

▲ Using fewer pesticides on the soil allows wild flowers to grow in the midst of cereal crops.

◄ Increasing numbers of farmers are rearing animals such as pigs on natural fodder crops, without using intensive farming methods.

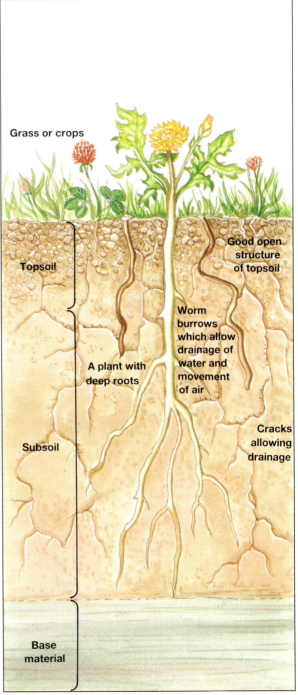

Grass or crops

Good open
structure
of topsoil

Topsoil

Worm
burrows
which allow
drainage of
water and
movement
of air

A plant with
deep roots

Cracks
allowing
drainage

Subsoil

Base
material

Using the same fields over and over again for the same crop tends to reduce the fertility of the soil and encourages pests and diseases. Organic farmers rely on **micro-organisms** in the soil and crop rotation to keep it healthy. Using crop rotation means that each field is sown with a different crop each year. This makes it more difficult for pests and organisms that cause disease to become established.

A section of soil showing good structure, root ▶
growth and earthworm activity.

Every few years the farmer plants a leguminous crop, which extracts nitrogen from the air and, when the crop is ploughed in, returns nitrogen to the soil. Alternating deep-rooted plants with shallow-rooted plants makes the best use of available minerals and helps break up the soil layers. If minerals are required, organic farmers use naturally occuring fertilizers, such as Chilean nitrate. Nothing, of course, is wasted. All animal wastes are returned to the land as manures and damaged crops are ploughed back into the ground. Unusable parts of crop plants are **composted** and then spread over the ground to enrich the soil.

The main disadvantage of organic farming is that food produced in this way is more expensive. Less food is produced per hectare on an organic farm than on a more conventional farm. As the human population increases, it is difficult to see how it would be possible to feed everyone by organic farming alone. However, supporters of organic farming claim that it is a solution to overproduction. There is a market for organically produced food, which consists of customers who are prepared to pay more for food produced organically. At present organic farming remains a relatively small part of the food-growing industry.

Organic vegetables are grown by gardeners as ▶ well as farmers, and some can be purchased in supermarkets or farm shops.

GLOSSARY

Altitude
The height of an object above the level of the sea.

Arable land
Land on which crops are grown.

Broiler houses
Chicken houses where large numbers of chickens are raised for eating.

Commerce
The business of producing and selling goods, such as crops and animals.

Compost
A mixture of vegetable matter and soil that is left to decay and then spread on the soil.

Extensive farm
Usually a large farm, where only a small amount of produce and income are obtained from each hectare of land.

Fallow
Refers to land that is either left uncultivated, or ploughed and harrowed but left unsown for a year.

Farrowing pens
Very small pens where sows give birth to their piglets. The pens are designed to stop the sows laying on their piglets.

Fertile
The ability of the soil to sustain a large amount of crops.

Fodder
Dried hay or root crops used for cattle feed.

Fungicide
A substance that destroys fungus.

Furrow
A trench made in the ground by a plough.

Germinate
When a seed starts to grow.

Hectare
A unit of measurement equivalent to 10,000 square metres or 2.471 acres.

Homoeopathic
Alternative medicine.

Hop
A climbing plant that produces cones which are added to beer to give it a bitter flavour.

Incubated
Eggs of animals or birds kept warm, to allow the young to hatch.

Intensive
Farming where a large amount of produce and income are obtained from every hectare of land.

Leguminous
Crops such as peas and beans whose seeds and root nodules can hold nitrogen.

Lochs
Long narrow bays or arms of the sea found in Scotland.

Micro-organisms
Tiny organisms that live in the soil and help keep it healthy.

Oast houses
Buildings containing ovens for drying hops.

Organic
The production of crops without the use of chemical fertilizers or pesticides.

Pastoral
Land used for pasture. Pastoral farming refers to the rearing of dairy and beef cattle and sheep.

Peat
A mixture of rotting vegetable matter and water. It is used for fuel and as a fertilizer.

Perchery birds
Egg-laying chickens kept in a house where they are free to roam and where perches are provided.

Silage
Winter feed for cattle, made from grass.

Soil erosion
The blowing away of soil by the wind.

Subsidies
Financial payments for farmers supplied by the European Parliament.

Surplus
More produce than is required.

Tubers
Fleshy underground stems of potatoes in which food is stored.

Uncultivable
Land that has poor soil and weather conditions making the growing of crops impossible.

BOOKS TO READ

Branwell, Martyn: *World Farming* (Usborne, 1994)

Flux, Paul: *Food and Farming* (John Murray, 1992)

Knapp, Brian: *Farms and the World's Food Supply* (Atlantic Europe Publishing Co, 1994)

Lambert, Mark: *Farming and the Environment* (Wayland, 1991)

INDEX

The statistics contained in this book are estimated figures and are taken from *Agriculture in the UK* (HMSO, 1994) and *Digest of Agricultural Census Statistics* (HMSO, 1993).